Samia Oueslati
Meriem Saada
Riadh Ksouri

Xerophytic enhancement: Asphodelus tenuifolius and Ononis natrix

Samia Oueslati
Meriem Saada
Riadh Ksouri

Xerophytic enhancement: Asphodelus tenuifolius and Ononis natrix

Phenolic composition, antioxidant power and antimicrobial activity

ScienciaScripts

Imprint

Any brand names and product names mentioned in this book are subject to trademark, brand or patent protection and are trademarks or registered trademarks of their respective holders. The use of brand names, product names, common names, trade names, product descriptions etc. even without a particular marking in this work is in no way to be construed to mean that such names may be regarded as unrestricted in respect of trademark and brand protection legislation and could thus be used by anyone.

Cover image: www.ingimage.com

This book is a translation from the original published under ISBN 978-620-6-71467-5.

Publisher:
Sciencia Scripts
is a trademark of
Dodo Books Indian Ocean Ltd. and OmniScriptum S.R.L publishing group

120 High Road, East Finchley, London, N2 9ED, United Kingdom
Str. Armeneasca 28/1, office 1, Chisinau MD-2012, Republic of Moldova, Europe
Printed at: see last page
ISBN: 978-620-7-75071-9

Table of contents

Table 1. Biological activities of phenolic compounds (Frankel et *al.,* 1995).

Table 2. Measurement of antibacterial activity of *Asphodelus tenuifolius* and *Ononis natrix* fractions using the disk diffusion method.

Table 3. Measurement of the antifungal activity of *Asphodelus tenuifolius* and *Ononis natrix* fractions using the disk diffusion method.

General introduction

The risks and harmful effects of synthetic antioxidants used as food additives have been highlighted in recent years (Bruneton, 2006), and the need to replace them with natural antioxidants has become inescapable. The latter represent an inexhaustible source of substances with a wide variety of biological and pharmacological activities. The scientific community is therefore increasingly interested in assessing their therapeutic properties, particularly those of plant-derived antioxidants. Indeed, plant-based treatments and a return to traditional medicine have come back to the fore, especially as phytotherapy offers remedies based on natural extracts that are well accepted by the body. In recent years, research into natural antioxidants has progressed (Prasad et al., 2009), including polyphenols, the richest class of antioxidants (Oueslati et al., 2012). These phenolic compounds are used in various applications in the food, cosmetics and pharmaceutical industries (Maisuthisakul et al., 2007). On the other hand, these compounds possess various biological properties including mainly anti-inflammatory, antimicrobial, anticancer and antiviral powers (Balasundram et al., 2007).

In addition, xerophilous spontaneous species are found on superficial soils, with dry pedoclimatic conditions linked to the sub-arid climate of the southern region, and are therefore extremophilous plants. These species are currently coveted for their richness in secondary metabolites, particularly phenolic compounds.

Indeed, since microorganisms have been identified as the cause of many diseases, particularly infectious diseases, scientists have begun to look for substances that could partially or totally inhibit their growth and development. On the other hand, a growing number of studies are showing the value of taking into account the interactions of vegetation with bacteria and fungi. To this end, we propose to study the variability of phenolic compound contents and antioxidant activities in two xerophilous plants, *Asphodelus tenuifolius* and *Ononis natrix*.

In the second part, an assessment of the antibacterial and antifungal potential of the various polar and apolar fractions was carried out.

Chapter 1: Bibliographical data

1. Phytotherapy and knowledge of medicinal plants

1.1. The history of phytotherapy

The history of phytotherapy is closely linked to that of modern medicine, which, since its discovery in the 19th century, has defined itself as a natural science. For this reason, it is normal and even necessary for the efficacy and safety of a therapeutic method derived from an ancestral tradition to be verified by modern scientific studies (Fintelmann and Weiss, 2000).

1.2. Definition

The term "phytotherapy" refers to the curative or preventive treatment of diseases and subjective disorders through the use of preparations obtained from whole plants and/or plant parts (leaves, flowers, roots, fruits, seeds). Plants used in this way are commonly known as medicinal plants. The plant or organ used for its medicinal properties always possesses several active ingredients. These must meet the requirements of drug legislation in terms of quality, efficacy and safety. With the use of isolated, chemically-defined plant constituents, we reach the limits of phytotherapy, which presents only one part of the study of medicinal plants, a vast science that also includes phytochemistry, phytopharmacy and phytopharmacology (Fintelmann and Weiss, 2000).

1.3. Specific action and general therapeutic effect

Each plant or plant part used has a therapeutic purpose, ranging from extended therapeutic action to specific, pharmaco-logically defined action, whether for crude extracts or isolated molecules. The more active ingredients a phytotherapeutic preparation contains, the greater its range of indications.

2. Description of the Liliaceae plant family

The Liliaceae family comprises almost 4000 species, present in all terrestrial ecosystems except those of very cold regions. They store their reserves in bulbs, rhizomes, roots and fleshy stems. It's a vast family, well represented in our region by numerous ornamental species (lilies, tulips...), food species (garlic, onion, leek...) and medicinal species (colchicum, ruscus...) . These plants are generally perennials with a bulb or rhizome. They are usually herbaceous, sometimes woody. They are typical monocotyledons: parallel venation, flower with six free or fused parts (3 sepals and 3 petals). (Reynaud Joël, 2001).

2.1. Asphodelus tenuifolius

2.1.1. Systematics

Kingdom: Plantae

Class: Liliopsida

Family: Liliaceae

Genus: *Asphodelus*

Species: *tenuifolius*

Common name: Berwagg

2.1.2. Botanical characteristics

Asphodelus tenuifolius is a large perennial plant, reaching 1 m in height (Figure 1.1). A relatively robust species, the developed flowering stem appears in spring and can reach up to 50 to 100 cm in length, with a branched tip. The flowers are hermaphroditic and white, visible from May onwards. The clustered inflorescences have a pinkish-white corolla. A bulbous, fleshy plant with a strongly tuberized root system (10 to 30 tubers), the leaves are light green in color, up to 20 to 30 cm long and around 3 cm wide (Chaieb and Boukhris).

Figure 1.1. *Asphodelus tenuifolius* from the Sfax region

2.1.3. Ecology and geographical location

(i) In Tunisia

It is a species with a fairly variable ecology, sometimes considered ruderal and nitrophilous (in rubble, cemeteries, etc.), or massicote (in tree and olive plantations). But unlike the forest formations of central and southern Tunisia, where this species is abundant, it appears only occasionally on the steppe. Its presence is reported in Sfax, Sidi Bouzid, Gafsa, and throughout the coastal zone as far as Djerba and Zarzis. It is absent from continental desert areas. (Chaieb et Boukhris).

(ii) In the world

Worldwide, *Asphodelus tenuifolius is found throughout the* Mediterranean, mainly in scrubland, and India, in the southern parts of France and Europe, along roadsides.

2.1.4. Interests

(i) Therapeutic interests

A. tenuifolius is of no pastoral interest, but has a wide range of therapeutic virtues. Its use is recommended for earaches, ulcers and breast abscesses. The roots seem to be used to extract a sticky substance (Chaieb and Boukhris). It is used for its diuretic, vulnerary and anti-spasmodic virtues. It is mainly used for various skin problems as a poultice on warts, wounds and eczema, reducing and dissolving swellings. It has an effect on menstrual obstructions and associated discomforts. In addition, the flower clusters are ornamental, and the bulb is used to treat scabies (a skin disease).

(ii) Maintenance interests

The roots of this plant were once eaten cooked with cereals or potatoes to make bread. When distilled, it produces a fairly strong alcohol. It is used as fodder for livestock, mainly sheep, and is a favorite with wild boar.

3. Description of the Fabaceae plant family

One of the largest angiosperm families, with almost 18,000 species worldwide, well represented in temperate regions. Fabaceae is the third largest phanerogam family (after Asteraceae and Orchids). The family owes its name to the ancient genus Faba, given to the broad bean (Michel Botineau, 1999). It's a family that includes species that are extremely important in the food (beans, peas, broad beans, etc.), agri-food (peanuts, soybeans, etc.) and agronomic (alfalfa, clovers, etc.) sectors. Many species are also of major ornamental interest (lupins, cytis, wisteria, broom...). Some are toxic (Reynaud

Joël, 2001).

3.1. *Ononis natrix*

3.1.1. Systematics

Kingdom: Plantae

Family: Fabaceae

Genre: *Ononis*

Species: *natrix*

Common name: Tfizza

French name: Bugrane coqsigrue

3.1.2. Botanical description

Champherophyte reaching 40 to 50 cm in height and highly branched. Branches are erect and slightly thorny. Leaves have 1 to 7 leaflets and developed stipules. The species is characterized by its intense spring flowering. The flowers, with their yellow corolla and well-developed vexillum (upper petal), appear in clusters along the stems. The pods are protruding and contain 3 to 7 seeds (Chaieb mohamed and Boukhris makki). This species mainly colonizes talwegs and streams.

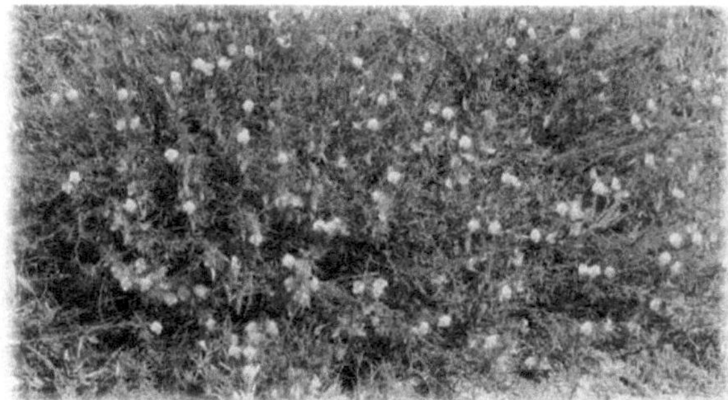

Figure 1.2. *Ononis natrix* in its natural biotope

3.1.1. Ecology and geographical location
(i) In Tunisia

In southern Tunisia, it exists in Sfax, Gafsa, Métlaoui, Djerba, Gabés, Médenine,

8

Bou Hedma, Tataouine and as far as Borj Bourguiba (Chaieb and Boukhris).

(ii) In the world

It is found in the Canary Islands, the Mediterranean, from Europe to Central Asia (Marion Dorr et *al.*, 2003). Stony or sandy limestone sites, almost all of France, southern Europe, western Asia and northern Africa (Chebli et *al.,* 2001).

3.1.2. Interests

(i) Interest in Q *natrix* seeds

A phytochemical study carried out on the seeds *of O. natrix, a* plant widespread in south-west Morocco, revealed its richness in major fatty acids, glycerides and flavonoids. Phytochemical selection revealed the presence of coumarins, catechic tannins, saponins and terpenes. GPC analysis of seed fatty acid methyl esters revealed the presence of linoleic and linolenic acids at 33 and 27 respectively. TLC analysis showed the presence of phospholipids and triglycerides. These seeds are also very rich in proanthocyanidins (prodelphinidin and procyanidin) and flavone aglycones, the most important of which are quercetin, kaempferol, isorhamnetin, nevadensin and penta-hydroxy-5,6,7,3',4'methoxy-3 flavone (Chebli et *al.*, 2001).

(ii) Therapeutic interests

Diuretic, antibacterial and antirheumatic properties have been attributed to extracts of the *Ononis natrix* plant (Muhammad et *al.*, 1998).

(iii) Nematicide benefits

Aqueous extracts of *O. natrix have been* tested for their nematicidal effect against root-knot nematodes (Meloidogyne ssp). Direct *in vitro* contact of nematodes with these aqueous extracts of this plant revealed a nematicidal power ranging from 60% to 95% (Chebli et *al.*, 2001). In parallel with this study, we carried out a phytochemical screening of this species to gain an overview of their secondary metabolite composition. Comparison of the efficacy of the aqueous extract with that of the commercial nematicide (Vydate) used as a control showed that *O. natrix* has a similar effect to the latter. This is probably due to the high alkaloid content of the plant's seeds (mainly harmine and harmaline).

4. The benefits of secondary metabolites

4.1. Definition of secondary metabolites

Secondary metabolites are non-enzymatic, low-molecular-weight antioxidants that form a second line of defense and include free-radical scavengers (koechlin-Ramonatxo., 2006).

4.2. Classification of phenolic compounds

Phenolic compounds can be classified into different categories, among which phenolic acids, flavonoids and tannins are considered to be the majority phenols in foods (Balasundram et *al.*, 2006), so these three major classes are identified according to the complexity of the basic molecule.

4.2.1. Phenolic acids

The term phenolic acid can be applied to all organic compounds with at least one carboxyl function and a phenolic hydroxyl. These phenolic acids make up around a third of dietary phenols, which can be present in plants in both free and bound forms. They share a common origin, phenylanin (Robbins, 2003). Phenolic acids can be divided into two groups: hydroxybenzoic acids and hydroxycinnamic acids (Manach et *al.*, 2004).

4.2.2. Flavonoids

The name flavonoid has been rather lent from flavus ; (flavus=yellow) (Malesev and kuntic, 2007). Flavonoids are the most represented group of plant phenols, accounting for more than half of the 8,000 phenolic compounds identified in nature (Heim. et *al.*, 2002). They are low-molecular-weight molecules consisting of 15 carbon atoms arranged in a C6-C6-C6 configuration (Ferrazzano et *al.*, 2011). They are rarely found in the free state and are almost always bound to sugars (Cheynier, 2006).All flavonoids share a common biosynthetic origin, phenylalanine. Flavonoids are widely found in plant leaves, seeds, bark and flowers, and are known to protect against ultraviolet radiation, pathogens and herbivores (Heim et *al.*, 2002).

4.2.3. Tannins

Tannins are the third most important group of phenolic compounds. They are high-molecular-weight secondary metabolites ranging from 500 to over 3,000 Da (Fogliani, 2002), and can be subdivided into hydrolysable and condensed tannins (Balasundram et *al.,* 2006). These compounds are partly responsible for organoleptic criteria such as bitterness and astringency (Dinnella et *al.*, 2011). They are the subject of a great deal of research focused in particular on understanding their structure, their nutritional role and the processes of their association with proteins (Falleh et *al.*, 2011).

Tannins are also thought to have antiviral properties, such as anti-HIV activity, due to their ability to reduce viral load by adhering to cell surface proteins, or their capacity to inhibit the enzymes essential for virus replication (Fogliani., 2002). Tannins are substances without nitrogen groups, they have an astringent taste and they color the skin through their binding to proteins (Fogliani., 2002). They also play an important role in determining the nutritional quality and organoleptic properties of food products derived from plants (Sun et *al.*, 1998).

4.3. Biological activities and industrial applications of phenolic compounds

Phenolic compounds found in plants are essential components of the human diet and are of considerable interest for their powerful antioxidant properties. In addition, polyphenols, including phenolic acids, flavonoids and tannins, have important biological properties such as anti-allergenic, anti-inflammatory, anti-thrombotic, antimutagenic and anti-carcinogenic, as well as cardioprotective and vasodilatory effects (Balasundram et *al*, 2006). In edible plants, these compounds act as antioxidant pigments (Erankel et *al.*, 1995). These substances are endowed with certain biological activities (Table 1.1).

Table 1. Biological activities of phenolic compounds (Frankel et *al.*, 1995).

Polyphenols	Biological activities	Authors
Phenolic acids such as benzoic acid	Antibacterial, antifungal and antioxidant	(Didry et *al.,* 1982) (Ravn et *al.,* 1984) (Hayase and Kato, 1984)
Coumarins	Protective of blood vessels and anti-edematous	(Makry and Ulubelen., 1980)
Flavonoids	Anti-tumor and anti inflammatory. Hypotensive, antidiuretic and antioxidant effects	(Stavric and Matula, 1992) (Das et *al.,* 1994) (Bidet et *al.,* 1987) (Bruneton., 1993) (Aruoma et *al.,* 1995)

Anthocyanins	Protection of the capillary-venous system	(Bruneton., 1993)
Proanthocyanidins	Anti-oxidant, anti-tumor, anti-fungal, anti-inflammatory. Stabilizing effects on collagen,	(Masquelier et al., 1979) (Bahorun et al., 1996) (DE oliveira et al., 1972) (Brownlee et al., 1992) (Kreofsky et al., 1992)
Gallic and catechic tannins	Anti-oxidants	(Okuda et al., 1983) (Okamura et al., 1993)

Phenolic compounds are also involved in the plant's relations with its biological environment (Heim et *al.,* 2002). For example, during pollination, insects are attracted to the colors of flowers by phenolic pigments. Phenolic compounds also protect plants from attack by pathogens such as insects, fungi, herbivores and bacteria (Heim et *al.*, 2002). Flavonoids protect plants against UV rays, pathogenic microbes and herbivores (Balasundram et *al.*, 2006).

5. Antimicrobial potential of secondary metabolites

Several *in vitro* and *in vivo* studies have focused on assessing the antimicrobial properties of secondary metabolites (Tepe et *al.*, 2004). Phenolic compounds, particularly flavonoids, inhibit bacterial growth and have a significant effect on various bacterial strains such as *Escherchia coli* and *Staphylococcus aureus* (Katarzyna et *al.*, 2007).

Antibacterial and antifungal properties of phenolic compounds against various strains have also been demonstrated (Gonçalves et *al.*, 2001). In addition, certain hydrolyzable tannins possess highly significant therapeutic properties, manifesting themselves in the inhibition of certain bacteria such as *Helicobacterpylori* (Funatogawa et *al.*, 2004). Various studies have demonstrated that the antimicrobial effect of secondary metabolites is due in part to a disruption of the lipid fractions of the microorganisms' plasma membrane, resulting in an alteration in the permeability of the microorganisms' plasma membrane. In addition, the physicochemical characteristics of phenolic compounds (water solubility and lipophilicity) can modulate this antibacterial effect (Domineco et *al.*, 2005).

5. Techniques for assessing antioxidant activity

Phenolic compounds are among the most powerful antioxidants, thanks to their ability to scavenge reactive oxygen species (ROS) and free radicals, as well as to reduce and chelate transition ions (Rice-Evans et *al.*, 1997). Several tests have been developed to assess this antioxidant capacity.

5.1. Total antioxidant activity

This activity involves testing the extract's antioxidant activity in its entirety, using the property of reducing Molybdenum (VI) to Molybdenum (V) and forming the Molybdenum (V) phosphate complex, which will be detected at 695 nm. Anti-oxidant activity is expressed in mg gallic acid equivalent per gram of dry matter (mg EAG $_g^{-1}$ MS) (Prieto *et al.*, 1999).

5.2. Anti-free radical activity (DPPH test)

This test enables us to assess this anti-free radical capacity using the spectrophotometric method, by monitoring the kinetics of the inhibition reaction through the discoloration of the synthetic free radical DPPH· (2-2-diphenyl-2-picrylhydrazine). The transition of this radical to a stable, non-radical form (DPPH-H) is achieved in the reaction medium by the attachment to (DPPH·) of a hydrogen proton released by the antioxidants in the plant extract (Galvez et *al.*, 2005).

5.3. Reducing power of iron

It's based on the redox reaction that takes place between the antioxidants (electron donors) in plant extracts and Fe^{3+} ions (electron acceptors). In fact, K_3 $Fe(CN)_6$ ferricyanidine (colorless) supplies Fe ions^{3+} which are reduced to Fe^{2+} by accepting an electron donated by the antioxidants. This test can be used to estimate reducing power by monitoring reduction kinetics spectrophotometrically (Gülcin et *al.*, 2006).

5.4. Chelating power

Chelating power refers to the plant extract's ability to chelate metal ions. Free metal (iron, for example) in the medium is stabilized by ferrozine, forming a ferrozine-metal complex (violet in color). Quantifying this complex spectrophotometrically provides information on the quantity of unchelated metal, and therefore on the ability of the plant extract to trap this metal (Rios et *al.,* 2006).

6. Techniques for assessing antimicrobial activity

In addition to their role as antioxidants, phenolic compounds have other properties, notably antibacterial and antifungal. In order to investigate the antimicrobial activities of these compounds on pathogens using plant extracts, we adopted the "Aromatogramme" technique, also known as the solid disk diffusion method. After culturing and diffusing the pathogens on these discs, plant extracts are added at different concentrations. Revelation is based on the observation of a halo around the disc, corresponding to the reduction in bacterial or yeast growth. The greater the diameter of this halo, the greater the antimicrobial activity (Rios et *al.,* 2005). These activities are compared with those of antibiotics as positive controls.

Chapter 2: Materials and methods

1. Sampling

The plant material collected in this study consists of two species: *Asphodelus tenuifolius* and *Ononis natrix*. Both species originate from the Sfax region.

Sampling carried out in February 2013 allowed us to sample the aerial parts of *Asphodelus tenuifolius* and *Ononis natrix* at the flowering stage.

Figure 2.1. Clump of *Asphodelus tenuifolius* (1) and *Ononis natrix* (2).

2. Extraction technique

The two species studied were immediately brought back to the laboratory for drying, first in the open air and then in an oven at 50°C for 72 hours. The aerial parts of both plants were then finely ground using a ball mill (Dongoumeau type). The powder was kept in the dark at 4°C for extraction and analysis.

- Methanol/chloroform/water extraction

A solvent mixture of methanol/chloroform/water (12 :5 :3 v/v) was used to separate hydrophilic from lipophilic compounds. A mass of 2.5 g of dry matter was added to 25 ml of the mixture (MCE). After 30 min maceration, the supernatant was recovered and a further 25 ml of solvent was added. This operation was repeated twice. The supernatants are collected and added with distilled water (V/V). To separate the aqueous phase (upper) from the apolar chloroform phase (lower), a decantation was performed. Both phases were evaporated under vacuum at 35°C using a rotary evaporator, until a dry extract was obtained. The dry residue was weighed and then taken up in pure methanol, to obtain respectively 10 mg/ml for the aqueous fraction (polar) and 2 mg/ml for the organic fraction (apolar). These two fractions are placed in a freezer at -20°C until the antioxidant activity biotests and/or purification of the active substances are carried out (Meot-Duros et al., 2009).

Figure 2.2. liquid-liquid extraction; polar and apolar phases

Figure 3.2. Evaporation of solvents using a rotary evaporator.

3. Determination and identification of phenolic compounds

3.1. Determination of total polyphenol content

Principle. In an alkaline medium, polyphenols reduce the Folin-Ciocalteu reagent to blue tungsten-molybdenum oxide. The intensity of this blue color provides information on the total polyphenol content of the mixture (Dewanto et *al.*, 2002).

Manipulation. 125µl of each suitably diluted extract is added to 500µl of distilled water and 125µl of Folin-Ciocalteu reagent. After instant stirring and a 3min rest, 1250 µl of $(Na)_2 CO_3$ (7%) is added, bringing the final volume to 3ml with distilled water. After 90 min incubation in the dark at room temperature, optical density was measured using a spectrophotometer at 760 nm. Three replicates were performed. The standard range is prepared with gallic acid at concentrations ranging from 50 to 500 mg.l^{-1} . Levels are expressed in mg of gallic acid equivalent per gram of dry matter (mg EAG.g^{-1} RS).

Figure 4.2. Determination of polyphenols and appearance of blue color

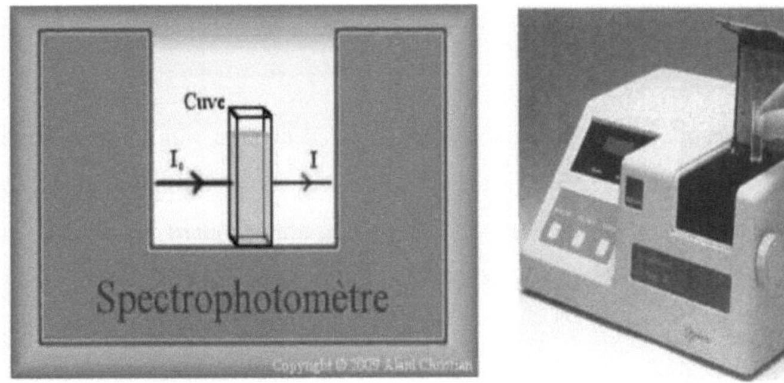

Figure 5.2. Schematic diagram of a cell displaying light flux and a spectrophotometer

3.2. Determination of flavonoid content

Principle. Flavonoids are quantified using a colorimetric method described by Dewanto et al. (2002). The intensity of the color (orange) indicates the flavonoid content of the plant.

Handling. 0.075 ml NaNO$_2$ (5%) was added to 0.25 ml of each extract. After a 6 min rest, 0.150 ml freshly prepared aluminum chloride (AlCl$_3$, 6H$_2$ O, 10%) was added to the mixture. After 5 min incubation at room temperature, 0.5 ml NaOH (1M) is added to the mixture and then made up to a final volume of 2.5 ml with distilled water. The absorbance reading was taken at a wavelength of 510 nm. The standard range is prepared with catechin at concentrations ranging from 50 to 400 µg.ml^{-1} . Flavonoid contents are expressed in mg catechin equivalent per gram of dry matter (mg EC. g^{-1} RS).

Figure 2.8. Flavonoid assay and appearance of orange color.

3.3. Determination of condensed tannin content

Principle. In the presence of sulfuric acid, condensed tannins depolymerize and react with vanillin to form red anthocyanidols, measurable spectrophotometrically at 500 nm (Sun et *al.,* 1998).

Handling. A 0.05 ml aliquot of the plant extract is added to 3 ml vanillin in methanol (4%) and 1.5 ml concentrated HCl. After 15 min incubation at room temperature, absorbance is measured at 500 nm. As with flavonoids, the standard is catechin, and condensed tannin contents are expressed as mg catechin equivalent per gram of dry matter (mg EC. g^{-1} RS).

4. Assessment of biological activities

4.1. Antioxidant activity

4.1.1. Total antioxidant capacity

This method involves adding 200 µl of plant extract to 2 ml of acid pH solution containing sulfuric acid (H_2SO_4 ; 0.6M), sodium phosphate (NaH_2PO_4 , H_2O; 28 mM) and ammonium heptamolybdate ((NH)$_{46}$ MO O_{724} , $4H_2O$); 4mM). The mixture is then placed in a water bath at 95°C for 90 min. After cooling to room temperature, absorbance is measured at 695nm. Total antioxidant activity is expressed as mg gallic acid equivalent per gram of dry matter (mg EAG.g^{-1} RS).

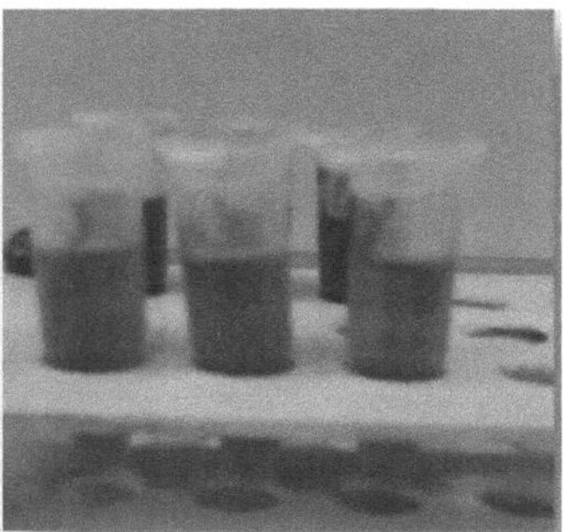

Figure 2.9. Total antioxidant capacity and appearance of the green-blue color

4.1.2. DPPH radical trapping

DPPH: 2,2'-diphenyl-1-picrylhydrazyl is a synthetic radical with an intense violet color in its oxidized state. The reduction of this molecule by protons from antioxidant substances induces the disappearance of the violet color, the degradation of which depends on the richness of the extract in molecules capable of trapping this radical. DPPH reduction highlights the anti-free radical power of the plant extract tested. Color degradation kinetics are determined by UV spectrophotometer at 517 nm, compared with a control (no extract) (equation 1).

$$\textbf{DPPH + AH} \longrightarrow \textbf{DPPH-H} \; + \; \textbf{A}^{\cdot} \qquad \textbf{(1)}$$

(violet) (Jaunâtre)

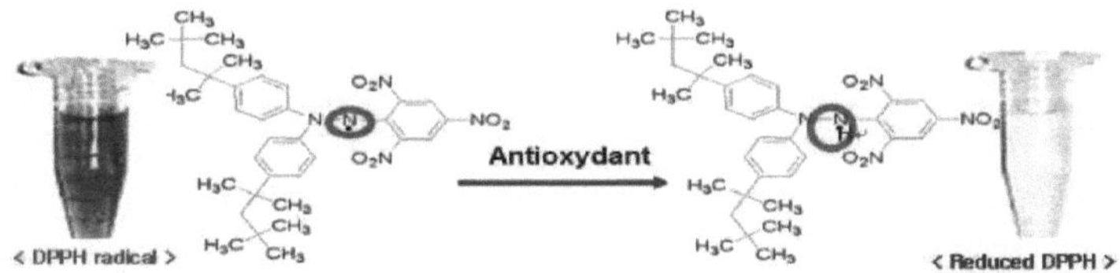

Figure 2.10. Degradation from violet to yellowish color

To measure this antiradicular activity, a 1 ml sample of the extract at different concentrations is placed in the presence of 250 µl of a DPPH· solution.

(0.2 mM in methanol). The mixture is incubated for 30 min in the dark, then absorbance is measured at 517 nm using a UV-visible spectrophotometer against a control (without extract). Results are expressed as percentages of inhibition, calculated from the decrease in color intensity of the mixture according to equation (2):

$$\textbf{PI = (DO témoin – DO extrait / DO témoin)*100} \qquad \textbf{(2)}$$

PI: percentage inhibition (or CI $)_{50}$

Control OD: absorbance of control

DO extract: absorbance of extract

By determining the kinetics of this activity, the concentration corresponding to 50% inhibition (IC_{50}) can be determined. The lowest IC_{50} value corresponds to the highest extract efficacy.

4.1.3. Measuring the reducing power of iron

Principle. The reducing activity is based on the redox reaction between the extract and transition metal ions, notably iron or copper (Huang et al., 2005). Indeed, $K_3\,Fe(CN)_6$ supplies $Fe^{3}+$ ions, which will be reduced by the ability of the plant extract's antioxidants to give up electrons according to equation (3):

$$\textbf{Extrait (antioxydants) + Fe}^{3+} \longrightarrow \textbf{Extrait (antioxydants) + Fe}^{2+} \qquad \textbf{(3)}$$

Handling. Reducing power is determined using the method described by Oyaizu (1986). This method involves mixing 0.2 ml of the extract at different concentrations (20 to 1000 $\mu g.ml^{\wedge 1}$) for the extract from the low-polarity phase and (1000 to 6000 $\mu g.ml^{\wedge 1}$) for the extract from the polar phase with 0.5 ml phosphate buffer (0.2 mol.l^{-1} , pH 6.6) and 0.5 ml potassium ferricyanide or K_3 Fe(CN)$_6$ (1%). The resulting mixture is incubated for 20 min at 50°C.

After this incubation, 0.5 ml trichloroacetic acid or TCA (10%) is added to stop the reaction, followed by centrifugation at *650xg* for 10 min at room temperature. To the 0.5 ml supernatant, 0.5 ml distilled water and 0.1 ml ferrous trichloride or $FeCl_3$ (0.1%) are added. Absorbance is read at 700 nm (the blank is the extraction buffer). Results are expressed as effective concentration (EC$_{50}$, µg.ml^{-1}), which is the extract concentration corresponding to an absorbance equal to 0.5. The EC value$_{50}$ is obtained from the linear regression curve (Mau et al., 2004).

4.1.4. β-carotdne bleaching inhibition test

This activity was determined according to the method of Kaur and Kapoor (2002), by dissolving 2 mg *^-carotene* in 20 ml chloroform. Then 4 ml of this solution were placed in a flat-bottomed flask with 40 mg linoleic acid and 400 mg Tween 40. After vacuum evaporation of the chloroform, the mixture is taken up in aerated distilled water. In a 96-well microplate, 150 µl of this emulsion is added to 10 µl of plant extract of known concentration. The microplates are then incubated at 50°C for 120 min and the OD is measured (at T=0 and T=120 min) at 470 min by a microplate reader (model EAR 400, Labsystems Multiskan MS). The activity of the extract is calculated in relation to that of the positive control, butylated hydroxyanisal (BHA), and that of the negative control (no extract). The percentage inhibition (PI) is obtained as follows:

$$PI = [(DO\ E_{120} - DO\ T_{120})/(DO\ T_0 - DO\ T_{120})]*100$$

OD E$_{120}$ = extract absorbance at t=120 min,

OD T$_{120}$ = absorbance of negative control at T = 120 min,

OD T$_0$ = absorbance of negative control at T = 0 min.

This activity is also expressed in CLo as described for the DPPH assay.

Figure 2.11. Placement of the prepared solution in the wells of the Elisa plate at T= 0 minutes

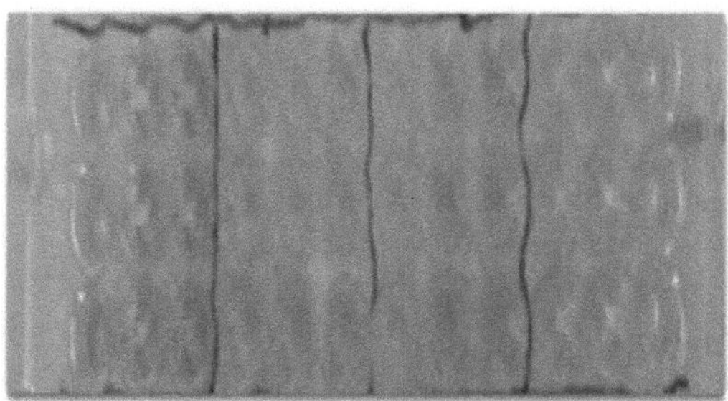

Figure 2.12. Reading via Elisa microplate reader.

4.2. Antimicrobial activity

4.2.1. Principle

The antibacterial activities of the two fractions (polar and slightly polar) obtained from *Asphodelus tenuifolius* and *Ononis natrix* against pathogenic bacteria were carried out using the disc (or diffusion) method, also known as the Aromatogram, based on diffusion on a solid agar medium (Graham et al., 1985). The choice of this method is justified by its reliability and simplicity (Rios et al., 2005).

4.2.2. Properties of bacterial and fungal strains

❖ **Bacterial strains**

The bacteria used throughout this work are: *Escherichia coli (ATCC 35218),*

Staphylococcus aureus (ATCC 29213), Micrococcus luteus (NCIB 8166) and *Enterococcus faecalis (ATCC 29212),* which are Gram-positive bacteria. *Pseudomonas aeruginosa (ATCC 27853), Klebsiella sp (CIP Tunis), Micrococcus luteus (NCIB 8166)* and *Salmonella thyphimurium (ATCC 14028)* are Gram-positive bacteria. *Shigella flexneri* belongs to the strict pathogenic Enterobacteriaceae, found exclusively in humans.

❖ Yeast strains

Candida albicans (ATCC 2091) and *Candida glabrata (ATCC 90030)* are non-pigmented yeasts. *Candida tropicalis 06-085* (Blastomycete of the order Cryptococcales, family Cryptococcaceae) and *Candida krusei (ATCC 6258)* (Hemiascomycete of the order Endomycetales, family Saccharomycetaceae) and finally *Candida krusei (ATCC 6258)* (Hemiascomycete of the order Endomycetales, family Saccharomycetaceae).

4.2.3. Preparation of culture media

Mueller Hinton medium. Mueller Hinton medium is a mixture of 15g agar, 17.5g hydrolyzed casein, 4g cattle meat infusion and 1.5g starch. 35g of this medium is hot dissolved in 1 liter of distilled water, with a final pH of 7.4. After sterilization by autoclaving for 15 min at 120°C, the medium is poured into Petri dishes at a rate of 15 ml per dish.

Sabouraud chloramphenicol medium. This medium is made from a mixture of 15 g agar, 10 g meat peptone, 20 g glucose and 0.5 g chloramphenicol. 42 g of this medium are dissolved, while hot, in one liter of distilled water, the final pH being 7. After sterilization by autoclaving for 15 min at 120°C, the medium is divided into Petri dishes at a rate of 15 ml per dish.

Inoculation by flooding. This stage begins with the preparation of a bacterial or fungal suspension. A pure, well-isolated strain is taken with a loop and dissolved in 10 ml sterile distilled water. The absorbance of this suspension is adjusted to 0.5 units at 570 nm. This mixture is then spread onto Petri dishes containing Mueller Hinton or Sabouraud chloramphenicol agar to almost completely cover the agar surface. Rotational movements imparted by the hand accelerate coverage. To fix the bacterial and/or yeast strains on the agar medium, incubate for 15 min at 37°C. The plates are then removed and the excess liquid sterilely discarded. A further incubation of the plates in the oven at 37°C for a quarter of an hour allows them to dry out and a bacterial or yeast mat to be obtained.

Disk application. Sterile 6 mm diameter wattman paper discs were placed on the bacterial and yeast mats. 10 µl samples of extracts prepared at a concentration of 30 mg ml^{-1} were then placed on the discs (Janssen et *al.,* 1987). After 24 hours incubation at 37°C, the effect of the extracts on the growth of strains around the disc is observed. The appearance of a halo around each disc indicates inhibition of bacterial and yeast growth (Baran et *al.,* 1994). This activity is measured by measuring the inhibition diameter. The positive controls used for these activities are antibiotics, in our case Gentamicin. t negative controls.

Chapter 3: Exploring the biological potential of Asphodelus tenuifolius and Ononis natrix

1. Introduction

The last decade has seen a resurgence of interest in the search for photochemical compounds from indigenous and natural plants for the pharmaceutical, cosmetics and food industries (Wangensteen et al., 2004). Thus, products of plant origin have great potential as a source of pharmaceuticals (Borchardt et al., 2008). This is mainly due to their high biological activities, sometimes exceeding those of synthetic antioxidants, which have a possible carcinogenesis-promoting activity (Suhaj, 2006). Antioxidants are also used to preserve food quality, mainly by stopping the oxidative degradation of lipids (Akinmoladun et *al.*, 2007). The evolution of antioxidant properties and free radical scavenging activities of fruits, vegetables and other plant products cannot be accurately achieved by any single, universal method or solvent extraction system due to the complex nature of the biomolecules present (Naczk and Shahidi, 2004; Prior et al., 2005).

Furthermore, it should be noted that the variability of phenolic compound content is generally correlated with the importance of biological activities. Moreover, antioxidant power is generally strongly correlated with phenolic compound concentration (Hanson et al., 2004). These data are corroborated by the work of Trabelsi et al. (2010), who showed a significant and positive correlation between phenolic compound content and free radical scavenging activity. However, this relationship is not always clear-cut, since it can be non-significant or even negative in some cases. Indeed, Djeridane et al (2006) consider that the existence of synergy between the various phenolic compounds may be a determining factor in the antioxidant capacity of a given plant. Thus, this activity depends not only on the polyphenol content, but also on the structure and interaction between the different compounds.

The aim of this chapter is to explore and compare the biological potential of the polar and apolar fractions of the two species *Asphodelus tenuifolius* and *Ononis natrix* by studying phenolic compound content (total polyphenol, flavonoid and condensed tannin content), investigating antioxidant capacity through various tests and finally estimating the antimicrobial potential of the different fractions.

2. Methodology

The aerial parts of two species, *Asphodelus tenuifolius* and *Ononis natrix,* collected from the Sfax region, were dried and then ground to a fine powder. The experimental protocol involved extraction with a mixture of MCE solvents and separation with distilled water to obtain two fractions (aqueous: polar and chloroformic: less polar), on which phenolic compounds such as total polyphenols, flavonoids and condensed tannins were assayed. The antioxidant potential of the various fractions was also determined.

The antimicrobial activity of both polar and apolar fractions was assessed using the agar disk diffusion method (Bagamboula et al., 2003). This involves inoculating bacterial and yeast strains onto Petri dishes containing the corresponding culture media to form bacterial and fungal mats. A sterile filter disk (Whatman paper no. 3) 6 mm in diameter was placed on the infusion agar seeded with bacteria, and a 10µl sample of a known concentration (30 mg.ml-1) of the extracts was dropped onto each paper disk. After this step, incubation at 37°C for 24 hours was carried out to examine the effect of these solutions on strain growth. Inhibition of this growth is shown by the appearance of an aureole around each disc; the larger the$_\chi$ diameter of this aureole, the greater the inhibition (Figure 3.2). Standard chloramphenicol disks (30µg) were used as positive controls. Bacteria studied included Gram-positive Cocci human pathogens such as *Staphylococcus aureus* (ATCC 25923), *Micrococcus luteus (NCIB 8166)* and *Enterococcus faecalis (ATCC 29212),* Gram-negative Cocci bacteria including *Escherichia coli (ATCC 35218),* Gram-negative bacilli such as *Pseudomonas aeruginosa (ATCC 27853), Salmonella typhimurium (ATCC 25922), Shigella flexneri* and *Klebsiella sp (CIP Tunis)* (Megdiche Ksouri Wided et al., 2011). Antifungal activity was summarized as previously described (Cox et al., 2000), the candida strains being *Candida albicans (ATCC 2091), Candida glabrata (ATCC 90030), Candida tropicalis (06-085)* and *Candida krusei (ATCC 6258).* Amphotericin B (10µg) was used as a positive control. Each experiment was performed in triplicate and the mean diameter of the zone of inhibition was recorded.

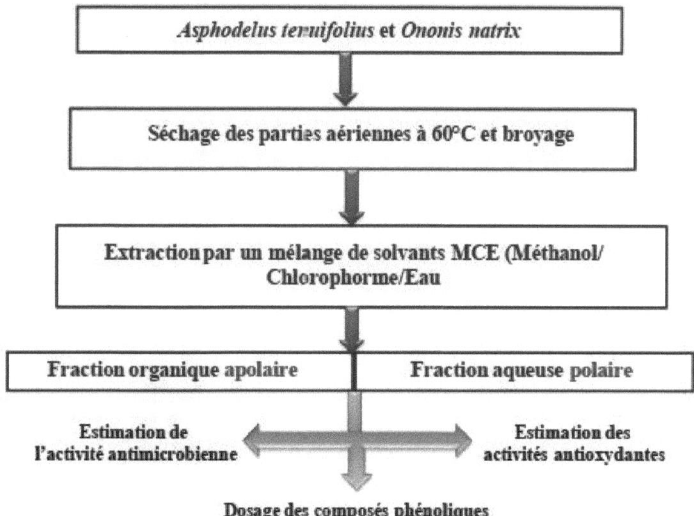

Figure 3.1. Experimental protocol for extraction using a mixture of extraction solvents performed on the aerial parts of *Asphodelus tenuifolius* and *Ononis natrix*.

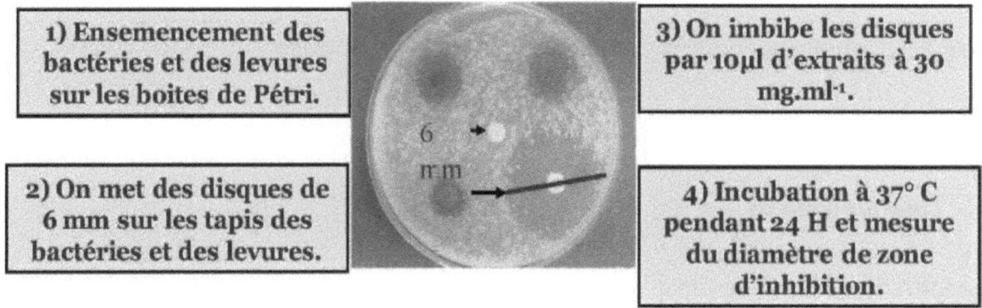

Figure 3.2. Experimental protocol for evaluating the antimicrobial activity of *Asphodelus tenuifolius* and *Ononis natrix* fractions against pathogenic bacteria and yeasts.

3. Results and discussion

3.1. Quantification of phenolic compounds in various fractions of the aerial parts of *Asphodelus tenuifolius* and *Oninis natris*

3.1.1. Total polyphenols

The results show remarkable variability in total polyphenol content between the two fractions (polar and apolar) and between the two species studied (Figure 3.3). Indeed, *A. tenuifolius* and *O. natrix* showed similar polyphenol contents (37 and 42.78 mg EAG. g^{-1} RS) in the polar fraction. Furthermore, *Ononis natrix* is the species with the highest levels of total polyphenols (142.3 mg EAG. g^{-1} RS) in the apolar fraction, 5 times higher than those found in *A. tenuifolius*.

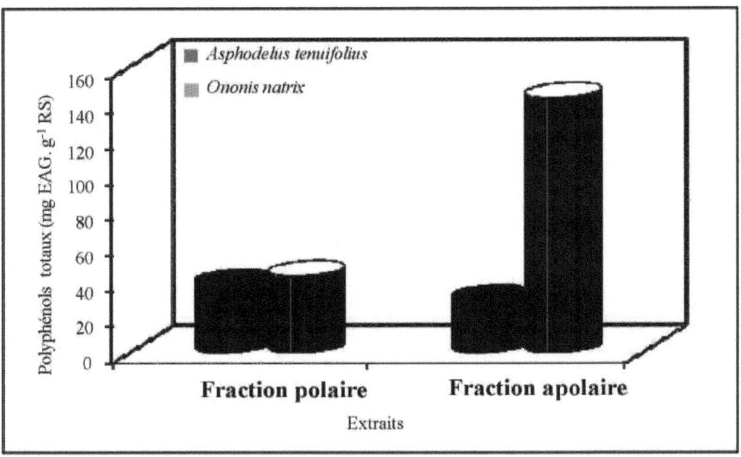

Figure 3.3. Polyphenol contents (mg EAG.g^{-1} RS) in the polar and apolar fractions of the aerial parts of *Asphodelus tenuifolius* and *Ononis natrix*.

3.1.2. Total flavonoids

Flavonoids are one of the most widely studied classes of phenolic compounds. Analysis of the results for the two fractions of the aerial parts of the two xerophytes shows a clear difference between the contents of these molecules in the two fractions, with the apolar phase showing higher flavonoid contents than those obtained in the polar phase. Moreover, as for polyphenols, *Ononis natrix* proved to be the richest species, with a content of 51.7 mg EC.g^{-1} RS in the chloroformic phase and 39 mg EC.g^{-1} RS. *Asphodelus tenuifolius, on the other* hand, had lower levels in both phases, at 16 and 28.3 mg EC.g^{-1} RS respectively (Figure 3.4).

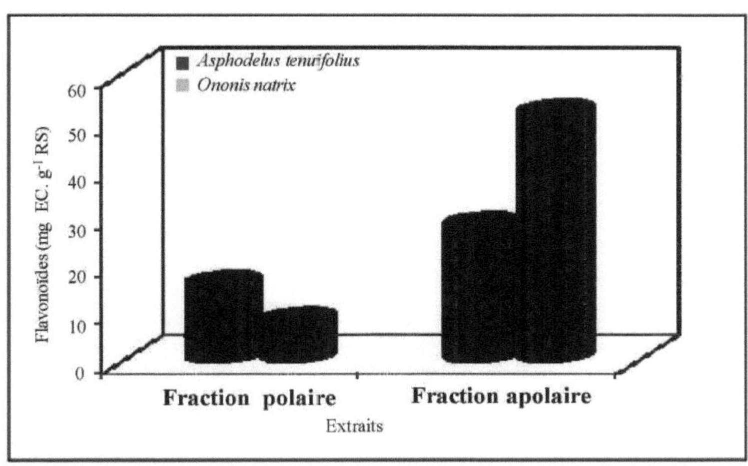

Figure 3.4. Flavonoid content (mg EC.g^{-1} RS) in the polar and apolar fractions of the aerial parts of *Asphodelus tenuifolius* and *Ononis natrix*.

2.2.3. Condensed tannins

Examination of the condensed tannin content of the different fractions confirms the variability in the content of these metabolites in the aerial parts of *Asphodelus tenuifolius* and *Ononis natrix,* depending on the two fractions (palar and apolar) (Figure 3.5). The apolar fractions of both species have high levels of condensed tannins, of the order of 20.7 and 27.5 mg EC.g^{-1} RS in *O. natrix* and *A. tenuifolius* respectively. As for the polar fraction, the levels measured are very low in both species. They are very close, at around 2 mg EC.g^{-1} for *A. tenuifolius* and 1.8 mg EC.g^{-1} RS for *O. natrix*.

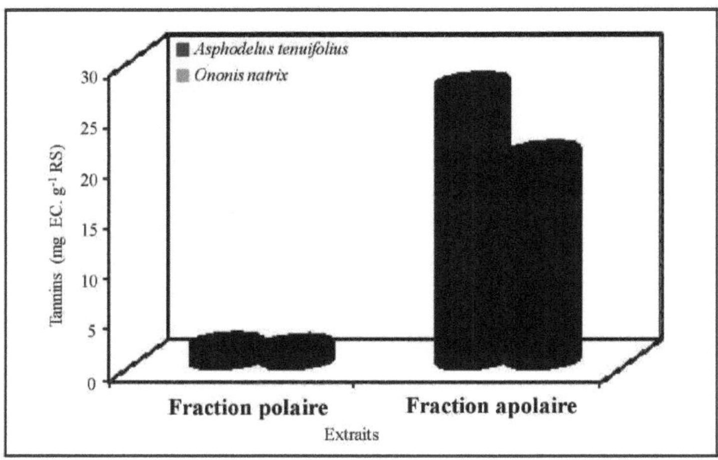

Figure 3.5. Condensed tannin content (mg EC.g^{-1} RS) in the polar and apolar fractions of the aerial parts of *Asphodelus tenuifolius* and *Ononis natrix.*

Based on phenolic compound content, *O. natrix* is more efficient than *A. tenuifolius*, with total polyphenol content reaching 142.3 mg EAG. g^{-1} RS. In addition, the polar fraction of this species is distinguished by its richness in secondary metabolites, particularly phenolic compounds. These results are in line with those of Turkmen et al. (2006), who showed that the nature and polarity of the solvent exerts a strong influence on phenolic extraction capacities in many species. Indeed, work carried out by Trabelsi et al. (2010) on the species *Limoniastrum monopetalum,* showed that total polyphenol contents using a polar solvent such as methanol were of the order of 15.8 mg EAG.g^{-1} MS, compared with those found using hexane (1.64 mg EAG.g^{-1} MS). In contrast, flavonoid and condensed tannin contents using water as the most polar solvent are significantly lower than those found using methanol. This highlights the influence of technical parameters such as the nature and polarity of the solvents used in optimizing the extraction of antioxidant molecules, particularly phenolic compounds. This could therefore explain the differences obtained in the quantities of bioactive molecules in *Asphodelus tenuifolius* and *Ononis natrix.*

3.2. Measurement of antioxidant activities of different fractions of the aerial parts

of Asphodelus tenuifolius and *Oninis natris 3 .2.1. Total antioxidant activity (AAT)*

Estimation of the total antioxidant activity of the various extracts of extremophilic species shows a clear difference between the two fractions (polar and apolar) and the two species studied. Indeed, *O.natrix* outperforms the other species as always, with a total antioxidant activity of around 101.5 mg EAG.g^{-1} RS for the polar fraction and 171.4 mg EAG.g^{-1} RS for the apolar fraction. In addition, *A. tenuifolius exhibits* moderately similar total antioxidant activity in both fractions (80.3 mg EAG.g^{-1} RS in the aqueous phase and 64 mg EAG.g^{-1} RS in the chloroformic phase respectively (Figure 3.6).

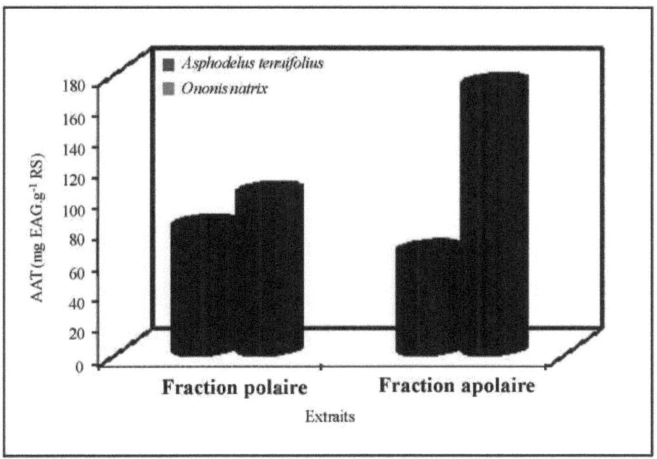

Figure 3.6. Total antioxidant activity (mg EAG.g^{-1} MS) of the two polar and apolar fractions of the aerial parts of *Asphodelus tenuifolius* and *Ononis natrix.*

3.2.2. Ability to neutralize the DPPH radical

The antiradical power of the aerial parts of *Asphodelus tenuifolius* and *Ononis natrix* was assessed by testing the inhibitory capacity of the synthetic radical DPPH at the 50% inhibition concentration (IC_{50}) (figure 3.7). The curves linking the various extract concentrations to DPPH degradation enabled us to determine the IC_{50}s corresponding to the fraction of each plant extract. Indeed, the polar fraction *of Asphodelus tenuifolius* shows a significantly more interesting antioxidant potential than that found by the apolar fraction where the IC50 is 400 and 10 µg.ml^{-1} , respectively. As for *Ononis natrix,* both fractions have the same capacity to trap the DPPH radical, with an IC50 of around 180 µg.ml^{-1} .

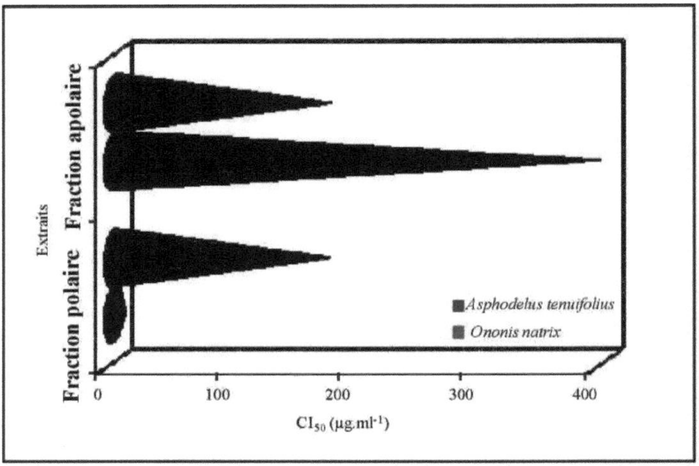

Figure 3.7. Antiradical activity (IC50 µg.ml^{-1}) of both polar and apolar fractions of extracts from the aerial parts of *Asphodelus tenuifolius and* Ononis natrix.

3.2.3. Reducing power of iron

Analysis of the results for the reduction capacity of transition metal ions, such as the ferric ion Fe^{3+} , shows a significant difference in this antioxidant activity depending on the polarity of the extraction solvent and the species studied. Indeed, the polar fraction *of Asphodelus tenuifolius stands out from* the apolar fraction for its high iron-reducing power, with an EC50 of around 500 µg.ml^{-1} versus 2200µg.ml^{-1} (Figure 3.8). *On the* other hand, *Ononis natrix* is the species with the lowest EC50 values, and therefore the most interesting, for both fractions (400 µg.ml^{-1} for the apolar fraction and 680 µg.ml^{-1} for the polar fraction).

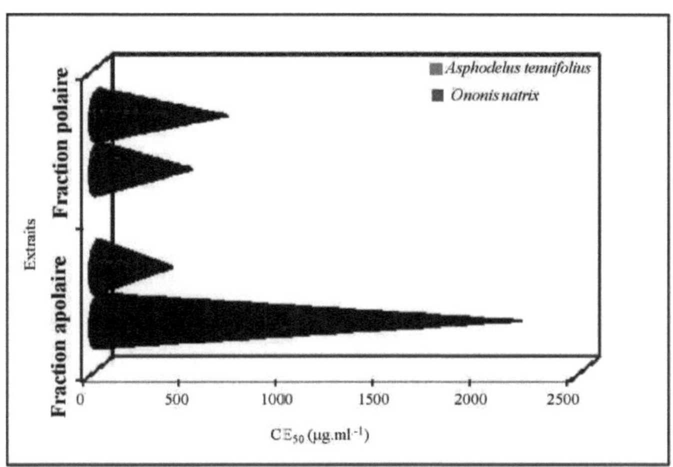

Figure 3.8. Iron-reducing activity (EC_{50} µg.ml^{-1}) in the polar and apolar fractions of extracts from the aerial parts of *Asphodelus tenuifolius* and *Ononis natrix.*

3.2.4. P-carotene bleach inhibition test

Estimation of the bleaching inhibition concentrations of β -carotene shows a difference between the two fractions and the two species studied. Indeed, *Asphodelus tenuifolius* is the species with the lowest IC value$_{50}$ for both fractions (1000 µg.ml^{-1} for the apolar fraction and 1600 µg.ml^{-1} for the polar fraction) (Figure 3.9). This ability to inhibit β -carotene bleaching is weaker in *Ononis natrix*, especially in the polar fraction, with an IC_{50} of around 4600 µg.ml^{-1} (Figure 3.9).

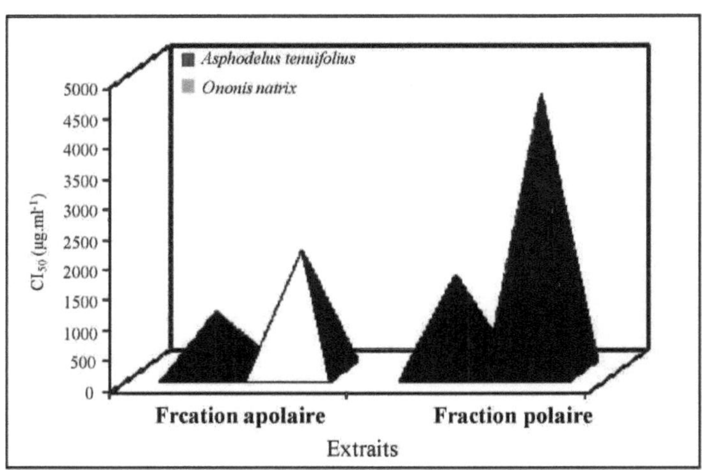

Figure 3.9. Bleaching capacity of β-carotdne (IC50 µg.ml^{-1}) in both polar and apolar fractions of extracts from the aerial parts of *Asphodelus tenuifolius* and *Ononis natrix.*

According to the results, antioxidant activity depends on the polarity of the solvent and the species studied. Indeed, the antioxidant capacity observed in the polar fraction of *Asphodelus tenuifolius* is more interesting than the apolar fraction of the same species. Unlike *Ononis natrix*, which exhibits superior antioxidant potential in the apolar fraction. This allows us to suggest the existence of a correlation between the total polyphenol content measured in the two fractions of the two species and antioxidant activity (Ksouri et al., 2008). In this context, Trabelsi et al (2012) showed that the high antioxidant power of *Limoniastrim guyonianum* stems and galls correlates with the condensed tannin content of these two organs. Thus, our results inspire that each activity translates into its own mechanism of action (Oueslati et al., 2012). In addition, we can say that the effectiveness of the activity in question is due to the presence of selective and specific antioxidant molecules (Saada, 2011). Indeed, the greater efficacy of a synthetic antioxidant such as BHT compared to those of our extracts of interest is most likely due to the fact that the latter is in its crude state and therefore purification of these antioxidant molecules can considerably improve their efficacy.

3.2. Antimicrobial activity (antibacterial and antifungal) of two fractions (polar and slightly polar) of *Asphodelus tenuifolius* and *Ononis natrix*

The antimicrobial properties of the various fractions were assessed using the agar well diffusion method against eight strains of Gram$^+$ and Gram$^-$ bacteria *(Enterococcus feacalis, Escherichia coli, Salmonella typhimirium, Staphylococcus aureus, pseudomonas aeruginosa, Klebsiella sp, Shigella flexenerii* and *Micrococcus luteus)* and against four *Candida (Candida albicans, Candida glabrata, Candida krusei* and *Candida tropicalis).* As with previous activities, antibacterial activity results varied between the two fractions and between the two species (Table 3.1). In fact, the inhibition zones vary from 6 to 12 mm. Thus, the apolar fraction of both species showed the best activity against all bacteria tested. Furthermore, the apolar fraction of *A. tenuifolius was* the most effective against the *Klebsiella sp* strain, with an inhibition zone diameter of over 11 mm. As for the apolar fraction of *Ononis natrix,* the best capacity was observed against *Staphylococcus aureus* and *Klebsiella sp* with an inhibition zone diameter equal to 12 mm. In addition. These two species were inactive against *Pseudomonas aeruginosa,* since the diameter of the zone of inhibition did not exceed 6 mm. In terms of antifungal activity, *A. tenuifolius* exhibited the best activity against *Candida albicans* and *Candida glabrata,* with an inhibition zone diameter of around 9 mm (Table 3.2). In addition, the apolar fraction of *Ononis natrix* is also active against *Candida glabrata,* with an inhibition zone diameter of 8.6 mm. This variability between fractions and between species depends on strain sensitivity.

Table 3.1: Measurement of the antibacterial activity of *Asphodelus tenuifolius* and *Ononis natrix* fractions using the disk diffusion method.

Bacteria	Diameter of inhibition zone (mm ± SD)				
	Aspliodelus renuifolius		Ononis narrix		
	F. polar	F.apolar	F. polar	F.apolar	Chloramphen. icol (30µg)
Escherichia coli A TCC 35218	7=0	7=0	7=0	7=0	34
Staphylococcu saureu s AT CC 29213	9.3=0.6	11=0	10.3=0.6	12=0	36
Pseudomonas UeruginosaATCC27853	6=0	6=0	6=0	6=0	12
Klebsiellasp. CIP Titnis	10=0	11.3=0.6	10.3=0.6	12=0	26
Micrococcus luteus NCIMB 8166	7=0	8.3 =0.6	7=0	8.3=0.6	16
Enterococcus faecalis A T CC 29212	10=0	10.6=0.6	10=0	11=0	46

Salmonella thyphimuriumATCC 14028	7=0	7=0	7=0	7=0	22
Shigella flexenerii ATC C 2 9903	7=0	8=1.1	7=0	9=0	16

Table 3.2. Measurement of the antifungal activity of *Asphodelus tenuifolius* and *Ononis natrix* fractions using the disk diffusion method.

Diameter of inhibition zone (mm ± SD)

Candida spp.	Asphodelus tennuifolius		Ononis natrix		AmphoteriCineB (10 µgml)
	F. polar	F.apolar	F. polar	F.apolar	
Candida albican s AT CC 2091	8.6±0.6	9±0	8.3±0.6	8±0	10.3±05
Candida glabrata ATCC 90030	9±0	8±0	7±0	8.6±0.6	9.6±0.5
Candida krusei ATCC 6258	7±0	7±0	7.3±0.6	7.6±0.6	12±0
Candida tropicalis 06-085	7±0	7±0	8±0	7±0	6±0

Since ancient times, many plants have been traditionally used to treat various illnesses. Currently, there is great progress in the use of these plants for the manufacture of antibiotics and drugs for the treatment of many infections caused by various pathogens (Karou et al., 2005). With the aim of *in vitro* research into the antimicrobial activities of natural compounds from plant extracts on pathogens, this test was carried out on different fractions of *A. tenuifolius* and *Ononis natrix*. The latter showed different capacities, in fact, the apolar fractions were more effective at inhibiting the growth of these pathogens than the polar fractions. These results suggest that the phenolic compounds in the apolar fractions are more potent on these pathogens. This ability is probably linked to the richness of these fractions in phenolic compounds, particularly in total polyphenols, whose levels exceed 171.4 mg EAG g^{-1} RS in *Ononis natrix*. Moreover, this activity depends on the sensitivity of the bacteria tested. In this context, several studies have shown a close relationship between the sensitivity of different bacteria (Gram-positive and Gram-negative) and the antibacterial efficacy of plant extracts (Rodríguez Vaquero et al., 2007).

General conclusion and outlook

This study was carried out on two xerophytes *Asphodelus tenuifolius* and *Ononis natrix* from the Sfax region. The study set the following objectives: (i) to investigate the phenolic compound richness of the polar and apolar fractions of these two species, (ii) to explore the variability of their antioxidant capacities via various tests, (iii) to determine the antimicrobial activity of these fractions.

The experimental work began by extracting the aerial parts of the two xerophytes using a mixture of solvents of different polarity, in order to estimate the extraction power of the different fractions obtained, and their influence on antioxidant activities. Analysis of the results of this first part revealed that, irrespective of the species studied, phenolic compound contents varied significantly according to the nature of the fraction and the polarity of the extraction solvents. Indeed, the apolar fraction *of Ononis natrix* showed the highest levels of polyphenols (171.4 mg EAG.g^{-1} RS) and flavonoids (51.7 mg EC.g^{-1} RS). On the other hand, the highest levels of proanthocyanins (27.5 mg EC.g^{-1} RS) were obtained in the apolar fraction of Asphodelus tenuifolius. This highlights the influence of technical parameters such as the nature and polarity of the solvents used in the extraction of antioxidant molecules, particularly phenolic compounds. Moreover, antioxidant potential varies according to solvent polarity and species. For example, the polar fraction *of Asphodelus tenuifolius* shows the best antioxidant activity, with the lowest IC$_{50}$ and EC$_{50}$ values. Conversely, the apolar fraction *of Ononis natrix* is characterized by the most interesting antioxidant capacities. Finally, the efficacy of these fractions against bacterial strains and yeasts was tested. The results showed that the apolar fractions of both species had interesting antibacterial activity, inhibiting the following pathogenic strains: *Enterococcus feacalis, Staphylococcus aureus, Klebsiella* sp, *Candida albicans* and *Candida glabrata*.

In short, the *Ononis natrix* species has distinguished itself as the xerophyte richest in natural antioxidants, giving it the highest antioxidant and antimicrobial capacities.

As prospects, it would be interesting to develop the following aspects:

- Explore other biological activities on *A. tenuifolius* and *O. natrix* fractions.
- Identify the phenolic compounds responsible for these biological activities

References

Anderson KJ, Teuber SS, Gobeille A, Cremin P, Waterhouse AL, Steinberg FM. 2001. Walnut polyphenolics inhibit in vitro human plasma and LDL oxidation.Biochemical and molecular action of nutrients. *J. Nutr.* 131: 2837-2842.

Akowuah GA., Ismail Z., Norhayati I., Sadikum A. 2005. The effect of different extraction solvents of varying polarities on polyphenols of *Orthosiphon stamineus* and evaluation of the free radical-scavenging activity. *Food Chem.* 93: 311-317.

Balasundram N., Sundram K., Samman S. 2006. Phenolic compounds in plants and agriindustrial by products: Antioxidant activity, occurrence, and potential uses. *Food Chem.* 99: 191-203.

Bruneton J. 2006. Pharmacognosie, Phytochimie, Plantes Médicinales, 3ème édition, 4ème tirage; Editions médicales internationales, Lavoisier, Paris.

Chaieb M., Boukhris M. 1998. Flore succincte et illustrée des zones arides et sahariennes de Tunisie. Association pour la protection de la nature et de l'environnement (Eds), Sfax. 98-102 pp.

Chebli B., Idrissi Hassani, Hmamouchi M. 2001- Acides gras et polyphénols des graines *dOnonis natrix* L. (Fabaceae) de la région d'Agadir,Maroc. 148: 333-340.

Closs B. 2002. Valorization of polyphenols in cosmetology. In: Polyphenols. Recent advances in polyphenols research. XXI International Conference on polyphenols. *Marrakech- (Morocco), University Cadi Ayyad, Faculty of Sciences Semlalia.* 186-191.

Cox SD, Mann CM, Markham JL, Bell HC, Gustafson JE, Warmington JR, Wyllie SG. 2000. The mode of antimicrobial action of the essential oil of *Melaleuca alternifolia* (tea tree oil). *J. Appl. Microbiol.* 88: 170-175.

Cheynier V., Duenas-Paton M., Salas E., Maury C., Souquet JM., Sarni-Manchado P., Fulcrand H. 2006. Structure and Properties of Wine Pigments and Tannins. *Am. J. Enol. Vitic.* 57: 298-305.

Dinnella C., Recchia A., Tuorila H., Monteleone E. 2011. Individual astringency responsiveness affects the acceptance of phenol-rich foods .*Appetit.* 56: 633-642.

Djeridane A., Yousfi M., Nadjemi B., Boutassouna D., Stocker P., Vidal N. 2006. Antioxidant activity of some Algerian medicinal plants extracts containing phenolic compounds. *J. Food chem.* 97: 654-660.

De Abreu I.N., Mazzafera P. 2005. Effect of water and temperature stress on the content of active constituents of *Hypericum brasilienne* choisy. *Plant Physiol. Biochem.* 43: 241-248.

Duros ML. 2009. Osmoregulation and production of active molecules in halophytes from the Breton coast. *University thesis, Brest.* 1-245.

Dewanto V., Wu X., Adom KK., Liu RH. 2002. Thermal processing enhances the nutritional value of tomatoes by increasing total antioxidant activity. *J. Agri. and Food Chem.* 50: 3010-3014.

Dixon RE., Hughes JM., Thorns berry., C. 1985. Disk diffusion antimicrobial susceptibility testing for clinical and epidemiologic purposes. *Am. J. Infect. Control.* 13: 241-249.

Dinis TCP., Madeira VMC., Almeida LM. 1994. Action of phenolic derivates (acetoaminophen, salycilate and 5-aminosalycilate) as inhibitors of membrane lipid peroxidation and as peroxyl radical scavengers. *Arch. Biochem. Biophys.* 315: 161-169.

Eckhard W., Marion D., Diego R., James N. 2003. Externally Accumulated Flavonoids in Three Mediterranean *Ononis* Species. 58: 771-775.

Espin JC., Garcia-Conesa MT., Tomas-Barberan FA. 2007. Nutraceuricals: facts and fiction. *Phytochem.* 68: 2986-3008.

Fintelmann V., Weiss RF. 2000. Practical manual of phytotherapy. 80: 3 -11.

Ferrazzano GF., Amato L., Ingenito A., Zarrelli A., Pinto G., Pollio A. 2011. Plant polyphenols and their anti-cariogenic properties: A review. Molecules. 16: 1486-1507.

Fogliani B., 2002. From physiological knowledge of *Cunoniaceae* endemic to New Caledonia, to research into the physicochemical and biological characteristics of their bioactive substances of interest. *Doctoral thesis in plant physiology and phytochemistry,*

42-52.

Fallah H., Oueslati S., Guyot S., Ben Dali A., Magné C., Abdelly C., Ksouri R. 2011. LC/ESI-MS/MS characterization of procyanidins and propelargonidins responsible for the strong antioxidant activity of the edible halophyte *Mesembryanthemum edule* L. *Food Chem.* 127: 1732-1738.

Frankel EN., Water house AL., Teissedre PL. 1995. *Agric. Food. Chem.* 43: 221-235.

Galvez M., Martin-Cordero C., Houghton PJ., Ayusom J. 2005. Antioxidant activity of methanol extracts obtained from *Plantago* species. *J. Agric. FoodChem,* 53: 1927-1933.

Gülcin I., Mshvildaze V., Gepdirmen A., Elias R. 2006. Screening of antiradical and antioxidant activity of monodesmosides and crude extract from *Leontice smirnowii* tuber. *Phytom.* 13: 343-351.

Graham DR., Dixon RE., Hughes JM., Thorns berry C. 1985. Disk diffusion antimicrobial susceptibility testing for clinical and epidemiologic purposes. *Am. J. Infect. Control.* 13; 241-249.

Heim KE, Tagliaferro AR, Bobilya DJ. 2002. Flavonoid antioxidants; chemistry,. metabolism and structure-activity relationships. *J. Nutr. Biochem.* 13: 572-584.

Halliwell B. 1994. Free radicals, antioxidants, and human disease: curiosity,cause,or consequence Lancet 344:721-724.

Hu C., Zhang Y., Kitts DD. 2000. Evaluation of antioxidant and prooxidant activities of banboo Phyllostachys nigra var. Henonis leaf extract in vitro, *J. Agric Food Chem.* 48: 31703176.

Huang D., Ou B., Prior RL. 2005. The chemistry behind antioxidant capacity assays. *J. Agric. Food Chem.* 53: 1841-1856.

Karou D., Dicko MH., Simpore J., Traore AS. 2005. Antioxidant and antimicrobial acities of polyphenols from ethnomedicinal plants of Burkina Faso. *Afr. J. Biotechnol,*4: 823-828.

Ksouri R., Megdiche W., Falleh H., Trabelsi N., Boulaaba M., Smaoui A., Abdelly C. 2008. Influence of biological, environmental and technical factors on phenolic content and antioxidant activities of Tunisian halophytes. *C. R. Biol.* 331: 865-873.

Ksouri R., Megdiche W., Debez A., Fallah H., Grignon C., Abdelly C. 2007. Salinity effects on polyphenol content and antioxidant activities in leaves of the halophyte *Cakile maritime. Plant Physiol. Biochem.* 45: 244-249.

Koechlin-Ramonatxo C. 2006. Oxygen, oxidative stress and antioxidant supplementation or a different aspect of nutrition in respiratory diseases. Clinical Nutrition and Metabolism. 20: 165-177.

Kaur C., Kapoor HC. 2002. Antioxidant activity and total phenolic content of some Asian vegetables. *Int. J. FoodSci. Technol.* 37: 153-161.

Koleva I., Teris A., Beek V,. Linssen H., Groot A., Lyuba N., Evstatieva. 2001. Screening of Plant Extracts for Antioxidant Activity: a Comparative Study on Three Testing Methods. *Phytochem Anal.* 13: 8-17.

Lisiewka Z., kmiecik W., Korus A. 2006. Content of vitamin C, carotenoids, chlorophylls and polyphenols in green parts of dill *(Anethum graveolens L.)* depending on plant height. *J. Food. Comp. Anal.* 19: 134-140.

Macheix JJ., Fleuriet A., Jay-Allemand C. 2005. Plant phenolic compounds, an example of secondary metabolists of economic importance. Press Polytechniques et Universitaires Romandes (Eds), Lausanne.

Maisuthisakul P., Suttajit M., Pongsawatmanit, R. 2007. Assessment of phenolic content and free radical-scavenging capacity of some Thai indigenous plants. *Food Chem.* 4: 14091418.

Manach C., Williamson G., Morand C., Scalbert A., Remesy C. 2004. Bioavailability and bioefficacy of polyphenols in humans. I. Review of 97 bioavailability studies. *Am. J. Clin. Nutr.* 81: 230-242.

Mau JL., Chang CN., Huang SJ., Chen CC. 2004. Antioxidant properties of methanolic extracts from *Grifola frondosa*, *Morchella esculenta* and *Termitomyces* *albuminosus mycelia. Food Chem.* 87: 111-118.

Macheix JJ., Sapis JC., Fleuriet A. 1991. Phenolic compounds and polyphenol oxidase in relation to browning in grapes and wines, *Crit. Rev. Food Sci. Nutr.* 30 : 441-486.

Michel Botineau. 1999. Systematic and applied botany of flowering plants. 80: 598.

Muhammad T, Saeed A., Nasim R., Abdul RR., Clive WB. 1998. Mass spectrometric studies of the principal dihydroisocoumarines of *Ononis natrix* and some related compounds.

Navarro JM., Flores P., Garrido C., Martinez V. 2006. Changes in the contents of antioxidant compounds in pepper fruits at different ripening stages, as affected by salinity. *FoodChem.* 96: 66-73.

Oueslati S., Ksouri R., Falleh H., Pichette A., Abdelly C., Legault J. 2012. Phenolic content, antioxidant, anti-inflammatory and anticancer activities of the edible halophyte *Suaeda fruticosa* Forssk. *Food Chem.* 132: 943-94.

Oyaizu M. 1986. Studies on products of browning reaction: Antioxidative activity of products of browning reaction. *Jap. J. Nutr.* 44: 307-315.

Pedras MSC, Ahiahonu PWK. 2005. Metabolism and detoxification of phytoalexins and analogs by phytopathogenic fungi. *Phytochem.* 66: 391-411.

Prasad KN., Hao J., Shi J., Liu T., Li J., Xiao W., Sheng XQ., Xue S., Jiang Y. 2009. Antioxidant and anticancer activities of high pressure-assisted extract of lon-gan *(Dimocarpus longan* Lour.) fruit pericarp. *Inn. FoodSci. Emerg. Technol.* 10: 413-419.

Prieto P., Pineda M., Aguilar M. 1999. Spectrophotometric quantitation of antioxidant capacity through the formation of a phosphomolybdenum complex: specific application to the determination of vitamin E. *Anal. Biochem.* 269: 337-341.

Reynaud Joël. 2001. La flore du pharmacien. Liliaceae. Fabaceae. 99-128 pp.

Rios JL., Recio MC. 2005. Medicinal plants and antimicrobial activity. *J. Ethnopharmacol.* 100:80-84.

Rice-Evans CA, Miller NJ, Paganga G. 1997. Antioxidant properties of phenolic compounds. *Trends Plant Sci.* 2: 152-159.

Rodríguez Vaquero MJ, Alberto MR, Manca de Nadra MC. 2007. Antibacterial effect of phenolic compounds from different wines. *Food Control.* 18: 93-101.

Saada M. 2011. Phytochemistry and biological activities of three medicinal species. Mastère Industries Alimentaires - Ecole Supérieure d'industrie Alimentaire à Tunis.80 p.

Sun B., Richardo-da-Silvia JM., Spranger I. 1998. Critical factors of vanillin assay for catechins and proanthocyanidins. *J. Agric. Food Chem.* 46: 4267-4274.

Shahidi F., Janitha PK., Wanasundara PD. 1992. Phenolic antioxidants, *Crit. Rev. Food Sci.* Nutr. 32: 317-157.

Trabelsi N., Megdiche W., Ksouri R., Falleh H., Oueslati S., Bourgou S., Hajlaoui H., Abdelly, C. 2010. Solvent effects on phenolic contents and biological activities of the halophyte *Limoniastrum monopetalum* leaves. *LWT.* 43: 632-639.

Trabelsi, N. (2012). Study of phenolic compounds and their biological activities in two local halophytes: *Limoniastrum monopetalum* L. and *Limoniastrum guyonianum* Boiss. PhD thesis in Biological Sciences. 157 p.

Tadhani MB., Patel VH., Subhash R. 2007. *In vitro* antioxidant activities of *Stevia rebaudiana* leaves and callus. *J. Food Comp. Anal.* 20: 323-329.

Toor RK, Savage GP, Lister CE. 2006. Seasonal variations in the antioxidant composition of greenhouse grown tomatoes. *J. Food. Comp. Anal.* 19 : 1-10. On the content of active constituents *of Hypericum brasilienne choisy. Plant Physiol. Biochem.* 43: 241-248.

Turkmen N., Sari F., Velioglu YS. 2006. Effects of extraction solvents on concentration and antioxidant activity of black and black mate tea polyphenols determined by ferroustartrate and Folin-Ciocalteu methods. *FoodChem.* 99: 835-841.

Printed by Books on Demand GmbH, Norderstedt / Germany